RETRACTED

The Man Who (Nearly) Faked His Way to a Nobel Prize

DYLAN KING

To all the people I've ever needed.

CONTENTS

PREFACE

Alfred Nobel, better known as the inventor of dynamite, was an absurdly wealthy Swedish businessman who died in 1895. In his will, he laid out his wishes to establish a yearly prize fund for the greatest human achievements in five categories: Physics, chemistry, medicine, literature, and world peace.

The Nobel prize isn't the end-all, be-all of human achievement. Plenty of other awards exist to celebrate the fields not covered by the six categories, not to mention the thousands of deserving nominations that don't go on to win. But there's some sort of innate satisfaction that comes from having this universally respected institution that can take a snapshot of our progress as a species, year after year.

In 1901, the first award in physics went to one guy who happened to discover something he called 'x-rays', and seven years ago a team of thousands measured gravitational waves from the collision of two black holes 1.3 billion years old.

Jan Hendrik Schön never won a Nobel prize. In fact, it's impossible to know if Hendrik Schön was even nominated.

Nominations for the prizes are kept secret, with only nominations older than 50 years being publicly listed. The most up-to-date version of the archive only includes nominations as recent as 1966, and even for just the first 65 years of the prize's existence, that's a total of 22,240 nominations. If Hendrik were to have been nominated, it most likely would have been in 2001 or 2002, which means we won't know for sure for another 30 years.

Of course, this isn't about a Nobel prize. In September of 2002, one month before the winners are traditionally announced, Jan Hendrik Schön is fired from his job. A little less than two years later, he'll lose his Ph.D. This is the story of the most blatant and egregious case of academic fraud in all modern physics.

1 PROFILE OF A WINNER

Jan Hendrik Schön was born in August of 1970 in northwestern Germany. As a young adult, he attended the Universität Konstanz on the German-Swiss border, and in 1993 he received the equivalent of a combination bachelor's and master's degree in the field of physics, with a specialization in electronics. As is common after graduation, Hendrik was offered to work on a Ph.D, which he would eventually graduate with in 1997.

Konstanz is a well-respected institution, and slackers who are just looking for a fancy piece of paper don't last long. As a student Hendrik was seen as a solid worker, but by no means as a genius. He was intelligent and had a keen sensitivity to other's results and expectations, but he was not brilliantly creative. When it came down to it, he would follow textbooks to the letter.[1]

He was quiet, contemplative, and when faced with a confrontation or disagreement he wouldn't put up much of a fuss. He was not the ambitious mind you'd normally associate with other famous innovators of science. This hardly paints the picture of a man who would take the academic world by storm in five short years.

So let me be clear: He got lucky. He got absurdly lucky. The reason Hendrik got any attention at all is because he just so happened to be at the right places, at the right times, with just the right people.

Hendrik's publications first gained significant traction in early 2000's and ramped up considerably throughout 2001. He wasn't just making headlines for a bunch of nerds, he was making headlines in the New York Times, so him getting the award in 2002 is well within the realm of possibility. Of course, he wouldn't be getting it alone.

Hendrik wasn't a lone wolf, and contrary to what we grew up learning, science is rarely pushed forward by lone visionaries. Unfortunately, the structure of the Nobel prizes encourages a false perception. In the 21st century, science is a far more collaborative endeavor. Controversially, they refuse to give out awards posthumously, leading to situations where key contributors who happen to have passed away are left as nothing more than footnotes in the history books. And, of course, awards can only be given out to one, two, or three individuals. Never more than three.[2]

This inevitably leads to some difficult and controversial choices as the years go on, as deciding how credit should be shared and between whom is hotly debated. Literature has almost exclusively been awarded to a single person, but the sciences have overwhelmingly been awarded to groups of two and three as the decades have marched on.

So yeah, Hendrik wouldn't be getting sole credit. He'd be sharing it with his team.

Jan Hendrik Schön: The wunderkind, the man with the magic touch.

Then you have Christian Kloc, the chemist.

Lastly, the one who lends the whole group legitimacy, Bertram Batlogg. The supervisor, a trusted name.

These three would have been a shoo-in for the 2002 prize. After all, they were co-authors on more than 70 of Hendrik's early papers.

Now, obviously the elephant in the room here is that you know the ending to this story— I already told you he gets caught. Surely the Nobel committee would know better than to nominate someone who would later come out as a fraudster?

But the Nobel committees— although we like to think of them as infallible— are still made up of just six (typically Swedish) humans. And humans have a history of (occasionally) making mistakes.

Take Enrico Fermi's prize from 1938. Now, Fermi, quite deservingly— is remembered as a giant of the field. He created the first nuclear reactor, theorized the neutrino, and wound up having an entire class of particles named after him.

But his Nobel prize— awarded for the creation of what we now call plutonium— turned out to be a mistake. He hadn't made plutonium at all; he had actually demonstrated nuclear fission.

His 'new' element was really a combination of smaller elements like barium and krypton.

You also have Johannes Fibijer's win in 1926 for medicine, when his theory that ringworms were causing cancer in rats was found to be incorrect. It was actually lack of Vitamin A that was causing the cancer. Of course, these two are rather benign examples.

The 1949 prize for medicine was awarded to Antonio Egaz Moniz, widely known as the father of the lobotomy; a procedure that was often administered without consent of the patient and is now illegal in most countries.

The Nobel committee is not without fault, and a premature award for falsified work would be far from the ceremonies' most controversial.

Do you remember how I mentioned Hendrik just happened to be at the right place at the right time?

That's because in the final year of his Ph.D thesis, his supervisor offered him an internship that would change his life forever.

So where was this internship?

Well, I've got a question for you.

What university do you think has the record for the most Nobel prize winning alumni, postdocs, and professors?

Go ahead, name a school. Better yet, name three schools. I guarantee you one of them breaks the top three.

Top of the list, with a commanding lead, is Harvard. Makes sense, it's well known for a reason.

The only one even in the same galaxy as Harvard is the University of Cambridge, then a little further down we have Berkeley.

Going further down than that, the spread becomes way denser. We've got Chicago, MIT, Columbia, Stanford, Caltech, Oxford, Princeton, Yale, all the other famous names that you grew up hearing as the dream schools for protagonists in high school dramas.

Now I have a much trickier question for you.

Ignoring all universities, ignoring all colleges, all publicly funded government institutions.
What commercial research lab do you think has the most Nobel prizes? That means which for profit, owned-by-a-big-business lab, whose job it is to make money?

I'm willing to bet 95 percent of you have no idea. Assuming you haven't looked at the section titles (please don't do that I'm trying to be dramatic).

There's a reason why all the top spots are universities. Universities are where passionate researchers go to study the questions that no business wants answered. If it's not a product, investors aren't going to fund it.

Nobel prizes are awarded to celebrate the massive strides humanity makes into the unknown, and the unknown isn't exactly something you can package up and sell at a Best Buy.

There is one major exception to the rule though, one place where the best and brightest minds were paid a salary to do research with no clear financial payoff.

This...

...is Bell Labs.

2 BELL LABS

These days, its' full title is Nokia Bell Labs, but it's gone by several other names and just as many owners throughout its history.

Bell Labs Innovations. AT&T Bell Laboratories. and first and foremost: Bell Telephone Laboratories. You may be thinking, 'wait, does that mean...?' Yeah, Bell Telephone Labs was named after none other than Alexander Graham Bell.

~~The inventor~~ The first man to patent the telephone. That's a whole scandal on its own, but probably best left for another time. He founded the American Bell Telephone Company in 1877, and through a bunch of convoluted business decisions to drive down their tax rate, eventually came to be known as American Telephone and Telegraph.

You know it as AT&T, which went on to found Bell Labs in 1925 as its Research and Development branch. After the telephone, Alexander Graham Bell himself had very limited involvement in the technology giant he gave birth to, and eventually fucked off to Nova Scotia to mess around with hydrofoils, solar panels, and.... eugenics!

Yep. Yep.

Now, as its many name changes suggest, Bell Labs very quickly grew beyond the narrow focus of telephones. After all, when your parent company controls almost all the telephone lines for the entirety of the United States and Canada, is the largest technological monopoly the world has ever seen, and was only brought down by the largest anti-trust lawsuit ever administered by the US Department of Justice, you tend to have a lot of money to throw at R&D.

15 employees of Bell Labs have won a combined 9 Nobel prizes. 8 in physics, one in chemistry. To put that in context, of the 213 laureates who've won for physics, 6.6 percent have some sort of tie to Bell Labs.

There's even a bit of a rounding error depending on how you count John Bardeen, who is one of only four people in history to win two Nobel prizes. These achievements range from math-intensive theoretical models to fundamental inventions that dragged us into the 21st century.

Did you know that it was two Bell Labs employees who recorded the first evidence of the Big Bang?

And you can thank two more for the invention of the CCD sensor, the first ever digital photo sensor that was essential to the creation of digital cameras.

Of course, the Nobel prizes fall short due to their focus on the natural sciences, and often underrepresent engineering and computer science.

I'd argue that many of Bell Labs most famous inventions, the ones you're most likely to recognize, are the ones that didn't win the Nobel prize. C, C++, and UNIX? They're all invented here. Stereo sound? Bell Labs. The decibel unit? Named after their founder. Solar cells, cell phone towers, the laser!

Okay, not actually the very, very first laser, but Ali Javan proposed and helped build the now iconic red helium neon laser.

I've left a ton of important milestones out, particularly in the realm of telecom and signal processing. The techniques they invented back in the telephone days have just continued to be scaled up into the era of fiber optics and are a large part of why your internet is measured in megabytes per second, and not kilobytes.

Bell Labs had become such a powerhouse of invention that by 1999, Bell Labs was applying for more than six patents a day and earning as many as four! There was, no joke, a digital patent clock in the front lobby of its new jersey HQ that kept ticking up day by day.[3] You know what? Here's every award they've ever won:

2018 Nobel Prize —Arthur Ashkin
2014 Nobel Prize —Eric Betzig
2009 Nobel Prize —Willard S Boyle and George E Smith
1998 Nobel Prize —Horst Störmer, Daniel Tsui, and Robert Laughlin
1997 Nobel Prize —Steven Chu
1978 Nobel Prize —Arno A Penzias and Robert W Wilson
1977 Nobel Prize —Philip W Anderson
1956 Nobel Prize —John Bardeen, Walter H Brattain and William Shockley
1937 Nobel Prize —Clinton J Davisson
2005 IEEE Medal of Honor —James L Flanagan
2001 IEEE Medal of Honor —Herwig Kogelnik
1994 IEEE Medal of Honor —Alfred Y Cho
1992 IEEE Medal of Honor —Amos E Joel, Jr
1989 IEEE Medal of Honor —C Kumar N Patel
1982 IEEE Medal of Honor —John Wilder Tukey
1981 IEEE Medal of Honor —Sidney Darlington
1980 IEEE Medal of Honor —William Shockley
1977 IEEE Medal of Honor —H Earl Vaughan
1975 IEEE Medal of Honor —John R Pierce
1973 IEEE Medal of Honor —Rudolf Kompfner
1971 IEEE Medal of Honor —John Bardeen
1967 IEEE Medal of Honor —Charles H Townes
1966 IEEE Medal of Honor —Claude E Shannon
1963 IEEE Medal of Honor —George C Southworth
1960 IEEE Medal of Honor —Harry Nyquist
1955 IEEE Medal of Honor —H T Friis
1949 IEEE Medal of Honor —Ralph Bown
1946 IEEE Medal of Honor —R V L Hartley
1940 IEEE Medal of Honor —Lloyd Espenschied
1936 IEEE Medal of Honor —G A Campbell

1926 IEEE Medal of Honor —G W Pickard
2014 US National Medal of Science —Shirley Ann Jackson
1996 US National Medal of Science —Chandra Kumar Naranbhai Patel
1996 US National Medal of Science —James L Flanagan
1993 US National Medal of Science —Alfred Y Cho
1991 US National Medal of Science —Arthur L Schawlow
1988 US National Medal of Science —William O Baker
1986 US National Medal of Science —Solomon J Buchsbaum
1982 US National Medal of Science —Philip W Anderson
1982 US National Medal of Science —Charles H Townes
1974 US National Medal of Science —Rudolf Kompfner
1973 US National Medal of Science —John Wilder Tukey
1966 US National Medal of Science —Claude E Shannon
1963 US National Medal of Science —John R Pierce
2010 US National Medal of Technology & Innovation —Michael F Tompsett
2006 US National Medal of Technology & Innovation —Herwig Kogelnik
2006 US National Medal of Technology & Innovation —James E West
2005 US National Medal of Technology & Innovation —Alfred Y Cho
2001 US National Medal of Technology & Innovation —Arun N Netravali
1998 US National Medal of Technology & Innovation —Dennis M Ritchie and Kenneth L Thompson
1994 US National Medal of Technology & Innovation —Richard F Frenkiel and Joel S Engel
1993 US National Medal of Technology & Innovation —Amos E Joel, Jr
1992 US National Medal of Technology & Innovation —W Lincoln Hawkins
1990 US National Medal of Technology & Innovation —John S Mayo
1985 US National Medal of Technology & Innovation —Bell Labs
2018 Draper Prize —Bjarne Stroustrup
2013 Draper Prize —Richard F Frenkiel and Joel S Engel
2006 Draper Prize —Willard S Boyle and George E Smith
1999 Draper Prize —John B MacChesney

1995 Draper Prize —John R Pierce
2011 Japan Prize —Dennis M Ritchie and Kenneth L Thompson
2003 Japan Prize —Seiji Ogawa 1985 Japan Prize —John R Pierce
2001 Kyoto Prize —Izuo Hayashi and Morton B Panish
1989 Kyoto Prize —Amos E Joel, Jr
1985 Kyoto Prize —Claude E Shannon
2017 Queen Elizabeth Prize —Michael F Tompsett
2017 Queen Elizabeth Prize —George E Smith
2017 C&C Prize —Alfred V Aho
2007 C&C Prize —John B MacChesney
1999 C&C Prize —George E Smith and Willard S Boyle
1997 C&C Prize —Barry G Haskell and Arun N Netravali
1995 C&C Prize —Alfred Y Cho
1995 C&C Prize —Akira Hasegawa
1991 C&C Prize —Jack M Sipress
1989 C&C Prize —Dennis M Ritchie and Kenneth L Thompson
1988 C&C Prize —Eric E Sumner, John S Mayo and M Robert Aaron
1986 C&C Prize —Morton B Panish and Izuo Hayashi
2017 Marconi Prize —Arun N Netravali
2009 Marconi Prize —Andrew Chraplyvy and Robert Tkach
2001 Marconi Prize —Herwig Kogelnik
1995 Marconi Prize —Jacob Ziv
1993 Marconi Prize —Izuo Hayashi
1992 Marconi Prize —James L Flanagan
1987 Marconi Prize —Robert W Lucky
1979 Marconi Prize —John R Pierce
1977 Marconi Prize —Arthur L Schawlow
2020 Turing Award —Alfred V Aho and Jeffrey Ullman
2018 Turing Award —Yann LeCun
1986 Turing Award —Robert E Tarjan
1983 Turing Award —Dennis M Ritchie and Kenneth L Thompson
1968 Turing Award —Richard W Hamming
1936 Academy Award —Edward C Wente and Bell Labs
2016 Grammy Award —Harvey Fletcher
2006 Grammy Award —Bell Labs
2020 Technology Emmy® Award —Nokia Bell Labs
2016 Technology Emmy® Award —Bell Labs
1997 Technology Emmy® Award —Bell Labs
2015 The Brain Prize —David Tank, Winfried Denk, Karel

Svoboda
    2014 World Technology Award —Holger Claussen
    2011 World Technology Award —Gabriel Charlet[4]

None of these are jokes. They've literally won an Oscar, two Grammys and three technology Emmys.

All of this is to say: if Bell Labs claimed they discovered something new, academia listened.

3 THE END OF MOORE'S LAW

I've left one invention for last though, because it's central to telling the rest of our story.

Bell Labs' first ever Nobel went to three men: John Bardeen, Walter Brattain, and William Shockley. It was in 1956 for the world-changing invention of the transistor. Their Nobel was a little belated, by almost a decade, but it was quickly becoming clear that the transistor had transformed the way we, as a society, were going to function.

You've almost certainly heard the word before: transistor. But what is it? What does it do?

Well, its most basic definition is 'a three terminal device used to amplify or switch electronic signals,' which I imagine doesn't clear much up. So, think of a two-terminal device, like a light bulb.

You hook it up to a battery, complete the circuit, now electrons can flow in and out of the bulb, heating up the filament, and producing light. Easy!

But what if we wanted to switch the flow on and

off?

We want to add in a valve. A third terminal that can shut the flow of electrons on and off. Of course, if it's just a light bulb we're talking about, simply add a light switch. A simple mechanical solution.

But what if you're making a wi-fi router? Like, if you wanted to send digital ones and zeros at gigabit speeds? Are you really going to flick that switch a billion times a second? See, a mechanical solution isn't going to cut it here. We need something way, way faster, and that's where the transistor comes in.

It's a device that turns off depending on whether the voltage you hook up is positive or negative. If current is flowing, then that's a 1. If current is blocked, then that's a 0. Without the transistor you can't add or subtract, without the transistor you can't multiply or divide. You don't have RAM, you don't have CPU's, you don't even have the video you're watching on YouTube to stave off sleep.

It is literally the most widely produced invention in the history of our species. Not the wheel, not the spear. We've made 13 sextillion of these bad boys! The modern world would not exist without transistors.

And it all started at Bell Labs.

In 1965, Gordon Moore made his iconic prediction: That the number of transistors that can fit on a chip would double approximately every two years which in turn would lead to a doubling of computing power. His informal 'law' has remained astonishingly accurate for the past 70 years, thanks to silicon.

Silicon is one of the group IV elements, which makes it a semiconductor. It sits right at the tipping point of the conductor/insulator spectrum, and a small push from a voltage will turn it on and off in the blink of an eye. The very first transistor was not made of silicon, it was made of germanium.

Both elements have their advantages and disadvantages but there's a key reason why silicon makes up 99.9% of all transistors produced since the 1950's: It's cost effective.

It scales well to mass production, and there are decades of infrastructure that has gone into foundries, clean rooms, R&D, and even trade deals. But there is an end in sight to Moore's law. At a certain point, shrinking your transistors down stops being feasible. You're working on the scale of individual atoms of silicon, and with so many atoms packed so tightly, the heating would fry your CPU.

Silicon just has a fundamental barrier at the nanoscale. So, what do we do? Rebooting the entire microelectronics industry with a new material is just unfeasible. It would have to be thousands of times cheaper to justify the changeover.

With no loss in performance, the silicon transistor fundamentally altered how the human race communicates, and it shapes the global economy to this day. When Moore's law fails, and it will, the economic ramifications will be immense.

But in the year 2000, something amazing happened.

An unknown German postdoc had unlocked a secret that could have saved Moore's law.

4 THE FUTURE IS PLASTIC

By the time the 90's rolled around, Bertram Batlogg was one of the most highly cited physicists in the world.

He was 4th in fact, out of a study of 1,120. With second and third place going to two other Bell Labs co-workers, because of course it did.

At the beginning of our story, he was the head of the Solid-State Physics and Material Research Division at the Bell Labs HQ in Murray Hill. Basically, the most pure-physics-y department in the whole company. More than anything, he was an expert on superconductivity.

That's the effect when you freeze material near to absolute zero and its conductivity becomes, well, super.

There's so many subtopics in physics that the field comes and goes in fads. A breakthrough emerges everyone scrambles to get in on the action. People gradually begin to realize they've exhausted all its potential and drop off, usually within the span of a decade.

Superconductors were one of those fads and it dominated the late 80s and early 90s. The holy grail was finding what you'd call a high-T superconductor. As phenomenal as superconductors were, they'd never be useful commercially because any money you'd save on power consumption would be wasted on cryostats to keep them refrigerated.

To this day, no one has ever found a room temperature superconductor, and the field is now left to the true diehards. Researchers at Bell Labs had historically been afforded a lot more freedom than the average scientist in the private sector, but Batlogg, even with his reputation, had to financially justify his research.

So, he pivoted in a new direction. After discussions with Bob Laudise, a senior director who would sadly pass away in 1998, the two of them came up with an idea:

Organic crystal semiconductors, A.K.A. plastic transistors. The word organic has varying meanings, when it comes to food, we tend to think pesticide-free, but when it comes to biologists, and especially physicists, it means only one thing:

Carbon.
Silicon's next-door neighbor.

But it's even more versatile when it comes to bonding with other elements and it's particularly good at forming infinitely long repeating chains called polymers, which in turn make up plastics.

Thing is, plastics are generally made up of massively chaotic tangles of carbon chains and rings. An electron floating through all that is not going to be easy.

They are— with the occasional exceptions— insulators.

Batlogg and Laudise hypothesized that if you take some of those organic molecules but grow them into large, ordered, neat and tidy crystals, electrons would have less to bump into and their conductivity would dramatically improve.

This made some intuitive sense. Silicon is grown in massive, ordered crystals, so why not organics? Even to physicists, though, crystal growth is thought of as somewhat of a black magic. You tinker around with different air pressures, gas combinations, humidities, and temperatures until you land on a recipe that works.

But with all that hassle, the advantage is that crystal growth is something you can do with a beaker of chemicals at your desk.

Growing silicon in mass scale quantities is expensive. It requires massive foundries and thousand-degree temperatures. Organics was worth giving a shot.

Christian Kloc had a decade of experience in crystal growth when he got recruited into Batlogg's research group in 1996, and he set to work refining his technique of growing large pristine crystals. But neither Batlogg nor Kloc had any experience doing semi-conductor measurements.

There was one missing piece of the team left to fill and so they reached out to Kloc's old lab at the Universität Konstanz, one that happened to be run by Hendrik's Ph.D supervisor.

Huh. Lucky for Hendrik.

Wrong!

He was even luckier than that!

His supervisor recommended two other students before him, but they had prior engagements.

And that was how Hendrik found himself with an internship at one of the world's most prestigious research labs.

Hendrik would spend January to May of 1997 in New Jersey, after which he went back to Konstanz to finish his Ph.D. His internship had gone well, and he returned to Bell Labs in the September of 1998 after spending most of the year waiting for a visa.

The difficulty with the sort of work that Batlogg's group was doing is that, for the majority of organic crystals, you're not going to see anything worth making a fuss over. Unexciting data that suggests 'hey, this specific crystal might be a dead end'. Well, that's necessary to the scientific process, even if it's not going to turn many heads. There's no shame in boring science, but it can be massively discouraging.

Between the start of his internship up until his eventual return as a postdoc, Hendrik struggled to get any results that were publishable, with some papers being outright rejected and others that were simply unremarkable. But this dry spell wouldn't last forever. In February of 2000, he'd finally done it.

Hendrik had done what had been considered impossible up to that point.

He'd found an organic crystal, Pentacene, with a conductivity that rivaled silicon.

Not only that, but he also managed to use pentacene to make two important building blocks of computer circuits. First, he made a transistor where the on/off current ratio was 108, a phenomenal result! He then used two transistors to make an inverter. If your input is high, your output is low. If it's low, then it goes high.

No matter how you looked at it, this was an indisputable breakthrough. This was leagues ahead of anyone else's work with organic crystals, but Hendrik had already gone ahead and made functional circuits out of it. After a frustrating start, Hendrik had found his footing in the cutthroat world of academia. This would have been a huge relief to Hendrik. By December of 2000, he had been upgraded from a temporary postdoc to a full-time salaried member of the technical staff.

At Bell Labs it was exceedingly rare for a postdoc to be offered a full-time job once the two-year term was up. One former Bell postdoc, who had hoped to get a permanent job in the exact same period as Schön, told me that publications strong enough to apply for professorships at good universities were not necessarily enough to get hired at Bell Labs in this period.[5]

Just like in academia, at Bell Labs it was publish or perish.

Hendrik chose the first option.

5 PUBLISH OR PERISH

What is a peer-reviewed study?

Like, I know you sort of know what it is, but what is it exactly?

Journals are private companies owned by traditional book publishers or subsidized by universities and science foundations.

The journals themselves don't do research. Scientists from relevant fields send in manuscripts according to the guidelines set out by the journals.
The journals employ editors who have backgrounds in the scientific fields.

They do a first screening of papers which filters out the bulk of submissions, and the ones they think are interesting and worthy of publishing they then send out to more experienced professionals in the field, working at universities or similar labs.

This is where the peer in "peer review" comes from.

The reviewers don't work for the journal. They're working for free and anonymously by writing valuable comments and critiques that they then send back to the editors, who then relay those back to the original researchers for changes.

This stressful tennis match goes back and forth for a few weeks to a couple of months and the paper either gets published or it gets rejected. Journals serve an important and necessary function by sharing scientific advancements with the rest of the world under a consistent set of standards. Publishing is not a quick process, nor is it an easy one.

It's absolutely not a perfect system though, don't get me wrong. At the end of the day, journals are still a business, which is why you usually need to shell out $15 a paper if you don't attend or work for a university.

With that in mind, let's pull up Hendrik's output. For the purposes of this, Scopus has the best analysis tools and it's especially accurate for anything post 1992, which most of Hendrik's work occurred after that.

Scopus lists 143 documents attributed to Jan Hendrik Schön.

Of those 143, 104 are standard articles: what you think of when you hear the words peer-reviewed study.

7 of them are conference papers: if you're going to present at a conference, you usually submit a written version of your presentation as well.

One of them is a review which isn't original research but rather a compilation that summarizes the state of the art in the field.

Another 31 are what we call an erratum, a fancy word for correction. We don't need to be looking at those because every single one is just a retraction notice.

To be clear: if 22 percent of your career history is corrections, something has gone terribly wrong.

The first publication ever to feature Hendrik as an author was during his Ph.D work in Konstanz, when he was working on materials for solar cells.

To this day, it only has 34 citations, and it didn't get its first one until two years after it was published. I gave it a read. It's fine. Like, I don't know man, it's not especially interesting. That's not the worst thing in the world, though. Plenty of people get Ph.D's on obscure subjects. Your Ph.D is meant to show that you can conduct independent research, not make you a celebrity.

As you can see here, the closer he got to the end of his Ph.D the more papers he started publishing, but nothing particularly earth-shattering.

Even after he began working for Bell Labs full-time, he still put out some papers with his old group in Germany and bingo, bango, bongo, you can see that he maintained a steady pace, but the larger scientific community still wasn't paying attention.

These columns account for 29 of those 104 articles I mentioned, so the remaining 69 were all produced at Bell Labs...

Huh. That's odd. Is that really all of them?

We still have 63 papers left. Are you trying to tell me that he put them all out in two years oh no oh no.

. . . .

In 2001, on average, he was putting out one paper every. eight. days.

I need you to understand that I do laundry less

frequently than that.

In December of 2001, he put out seven papers, which is almost two per week. The fact he was putting out this many papers was enough to raise some eyebrows, but it went beyond just his pace.

He was in the big leagues. Much like how there's a pecking order to traditional news outlets, the same can be said for scientific journals.

A good rule of thumb is that if a journal's name is one noun, it's probably a big deal.

Nature and Science are two of the oldest journals still publishing today, dating back to the late 1800's, Nature being based in the UK and Science in the US. Historically, they both had a bigger emphasis on biology papers, but by the 1990's both journals were making an effort to attract more physicists.

Who would you rather publish with, Solar Energy Materials and Solar Cells? Or Science?

I bring this up because Nature and Science are notoriously hard to get into. To anyone in research, a single one would be a career milestone.

Hendrik had done it 16 times and he was barely in

his 30s. How had he done it?

When it comes to the "big two", plenty of well-researched and thorough papers still get rejected. Although framed like non-profits, both journals generate significant revenue from ads. It's in their best interest to publish only the most exciting and novel discoveries.

Not only do you need to meet quality control, but your research can't just be incremental. It needs to push the field forward in a major way.

And let me tell you, no one in physics was doing anything half as exciting as Hendrik was.

6 WELCOME TO THE CLUB

His paper on pentacene transistors appeared in Science and it quickly caught the attention of the solid-state physics community. By 2002, it had been cited 194 times, putting it within the top 0.01 percent of all physics papers of the year 2000.

That might not sound like a lot, especially when we're used to seeing numbers in the millions for retweets and subscriber counts, but think about it like this:

Each of those citations is from an original piece of research carried out by teams of people that took months, maybe even years to carry out, write up, and then also get through the journal review process.

Hendrik had broken down the organics barrier. There was now a viable semiconductor to rival silicon and every lab in the country wanted in on the action. To this day, this is still Hendrik's most cited paper, and he took advantage of the hype to pump out several more papers on whatever organic crystal his buddy Kloc had been playing with that week.

But Hendrik's seemed to recognize that swapping out pentacene for a new organic crystal was going to get old at some point and the community would lose interest. So, he went several steps beyond. He had invented the world's first organic superconductor. Not just that, a high-T superconductor.

By high-T, we mean about 52 kelvin, so you know, not room temperature, but compared to zero kelvin, that is tropical. Thanks to one man, organics have gone from insulators with questionable utility to viable semiconductors, after which it completely skipped over conductors and hopped into the realm of superconductivity.

This was, as one researcher described it, like "turning an apple into an orange". At one point, a colleague told Batlogg, "You're going to put chemists out of a job," and Batlogg didn't deny it.[6]

Hendrik went wild with superconductivity. It seemed like every new crystal he touched had untapped potential no one had seen before. He didn't stop there though. Not long after he published a paper on the quantum hall effect.

The quantum hall effect is an obscure phenomenon where the resistivity climbs in a discrete staircase effect. It's observed only in temperatures close to absolute zero because you've removed all the thermal noise from the system and you're quite literally observing the flow of individual electrons.

Again, by demonstrating it in organics, Hendrik had claimed to another first for physics.

When Hendrik signed on to Bell Labs, Batlogg was already getting up in years. By the turn of the century, he opted to take a professorship at his alma mater, the Swiss institute for technology. Batlogg had been the trio's resident expert on superconductivity, and without his mentor to go to bat for him at conferences, Hendrik had to diversify his work.

He'd already mastered transistors and superconductors had apparently been a cakewalk.

What other flashy areas could he dip his toes into?

Well, uh... Here he is inventing the world's first and only organically driven laser. His second most cited paper.

Do you understand now?

Do you understand how batshit this man's output was?

He was a god.

On their own, transistors, superconductors, lasers, and the quantum hall effect all won Nobel prizes, sometimes multiple.

Over a span of two years, Hendrik released cover after cover of the greatest hits of physics, and his signature hook was doing them with organics.

After Batlogg had left Bell Labs for good, Hendrik continued to work on the crystals provided to him by Kloc and began collaborating with some chemists in France as well. But it was his collaboration with Bell Labs chemist Zhenan Bao which would prove to be his boldest— but also final— performance.

On paper, you can just make transistors smaller and smaller, pack more and more into the same space and computing power should keep increasing. But, in reality, silicon can't be miniaturized past a certain point.

Power density doesn't scale down alongside the transistor so you're still pumping the same number of watts per unit area, and with your transistors packed so tightly you get current leakage and overheating. It means at our current pace with silicon, Moore's law is going to fail.

But is there another way? What if instead of taking a piece of bulk material and shrinking that down, you could instead build a transistor from the ground up? Maybe even atom by atom?

Could we make a molecular transistor? This would be hundreds of times smaller than anything that's been done of silicon. It would be on the level of individual electron hopping. On paper, this is the true final barrier for Moore's law.

And in October of 2001, it looked like Hendrik and his partner Zhenan Bao had actually done it.

But is it Nobel prize worthy? That's the real question, right?

In truth, I'd argue that even without the crown jewel of his molecular transistor, Hendrik was well on his way to a Nobel prize, but it would be nice if we could back that up with something more concrete. It's one thing to boast about citations, but citations don't pay the bills.

In July 2000, along with Batlogg and Kloc, he shared a $10,000 industrial award at the International Conference of Synthetic Materials.

In October of 2001, they shared again the Braunschweig award for 100,000 DEM (~$70,000). After Batlogg moved into his professor role in Zurich, Hendrik began to emerge from his mentor's shadow.

Science mentioned his molecular transistor as part of their breakthrough of the year for 2001.

At the Materials Research Society spring meeting in San Francisco, he accepted the outstanding young investigator award, a $5,000 prize.

He made MIT's list of innovators under 35.

The soft-spoken Schön recalls being very surprised by how well his molecular transistors worked.[7]

Yeah, I bet he was.

In December of 2001, he traveled to Berlin to receive the Otto Klung Weberbank prize for 50,000 DEM (~$35,000).

The man handing him the novelty oversized check? That's Horst Stormer, who at the time was Bell Lab's most recent Nobel laureate.

He got his for the discovery of the fractional quantum hall effect, an effect Hendrik had just recently demonstrated in organics.

Turns out Horst Stormer had won the same prize as Hendrik in 1985 and later won his Nobel prize in 1998. The parallels between the two were pretty apparent, with a German headline reading "tipped for a Nobel prize".

Stormer is even quoted as telling Hendrik: "Welcome to the club".[8]

It's hard to trace rumors 20 years after the fact, but if I had to guess where the buzz around Hendrik winning the Nobel came from, I would guess here.

In 2002, it seemed like Hendrik was considering leaving Bell Labs behind. As well as being under consideration for a professorship at Princeton, he was near to getting an offer from the Max Planck Institute back in Germany.

Hendrik would have been the youngest person ever to receive a full research directorship there.

It would later come out that Hendrik had been chosen as the winner of the William L. McMillan prize, worth $2,500, which is given out by the

University of Illinois to a physicist within five years of finishing their Ph.D.

Hendrik had been on the shortlist for 2001, but they opted for someone else since his results had yet to be replicated. But a year later, the committee reasoned that even if half of his work turned out to be wrong, he was still ahead of the runner-up candidate.

Hendrik would never receive that prize letter. In May of 2002, the contest had been put on hold.

That same month, an editor from Nature had sent him an email asking for a clarification on what appeared to be a simple mix-up.

They just needed to know why these two graphs...

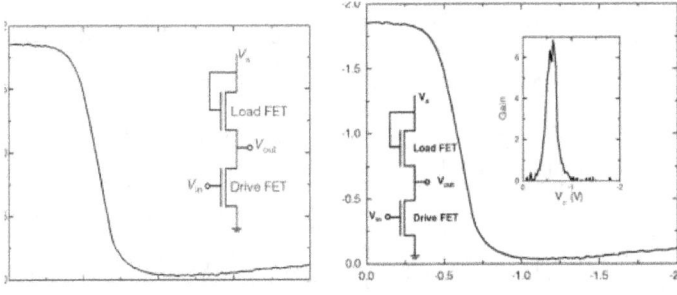

looked so similar.

Except this was no mix-up.

It looks like Bell Labs had a whistleblower.

7 DOUBLE BUBBLE

By the end of this, you're going to ask why he wasn't caught earlier. Every time I look at the evidence, I find it hysterical. It's just so laughably out in the open.

Who let this go unchecked for so long? Was no one watching? Was no one asking him the tough questions?

Was everyone too blinded by the glory that they hitched their cart to him and hoped for the best?

Partially, yes, but if the walls around you were on fire and a bucket of water showed up out of nowhere... Would you question how it got there, or would you use the bucket?

On March 10th, 2000, the NASDAQ index reached its then-peak of 5,048. A peak it wouldn't hit again until 2015. You've heard of the dot-com bubble, right? At the turn of the millennium, investments in internet businesses were at an all-time high. However, when it became clear that many of these kinds of silly businesses were never going to turn a profit, the market crashed.

Here's the part you probably didn't know: A year later in 2001, there was another crash. The telecom crash. It was 10 times bigger than the dot-com bubble.

See, the internet doesn't just run on its own, you need hardware to power the internet. You need millions of switches and thousands of kilometers of fiber-optic cable. And when the Clinton administration passed the telecommunications act of 1996, companies invested billions of dollars, most of it financed by debt (don't ask me how that works), so they could be part of the world's 3G infrastructure.

But when the dot-com bubble burst, which provided much of the funding for these networks, the telecom industry imploded. Companies like Cisco, Intel and Oracle lost 80 percent of their market value. And believe it or not, those were the lucky ones.

If you weren't declaring bankruptcy, you were laying off the equivalent of a mid-sized town. That may sound like I'm exaggerating but a company from Ottawa, JDS Uniphase, went from employing 10,000 workers to 580.

Lucent Technologies, who owned Bell Labs at the time, happened to be one of the lucky ones. In one day, Lucent's stock fell by 30 percent. Between 2000 and 2002, Lucent's stock had fallen by a factor of 40.

In the years prior to the crash, Lucent had spent a decent portion of its profits buying up other smaller competing companies. As soon as the crash hit, it did the opposite. Splitting up into spin-offs like Avaya Agere systems, or outright selling entire departments to competitors.

Hendrik's department was one of the few to stay on with Bell Labs. It was life or death for the remaining researchers. Half of the light bulbs were literally being removed from sockets to save money on power. And the beloved technical library was slated to close, much to the frustration of the staff.

Joke memos circulated about how printing with any font larger than 12 points was forbidden. Older employees who tended to have the largest salaries were offered voluntary retirement packages. In Hendrik's department, over the course of four years, the number of researchers fell from 114 to 56.

In the middle of all this, 9/11 happened within driving distance of Murray Hill, New Jersey.

You could definitely say that morale was not the highest at Bell Labs.

This was followed a month later by a near complete shake-up of the executive board of Lucent. Each and every department head now had to make a case to investors why they shouldn't have their funding cut. If you couldn't find a way to turn your project into a product, you might be out of a job.

The materials science division at Bell Labs was perhaps the most physics-y out of all the departments. The research they did there was important and fundamental to discoveries down the line but unlikely to lead to a new product anytime soon.

If any department was first on the chopping block, it was that one.

But if it had a breakout star who was making breakthroughs around the clock? Hell yeah!

Why wouldn't Bell Labs want to publicize that?

Hendrik was a bucket of water in an inferno.

8 BEST LISTENER IN PHYSICS

By all accounts, Hendrik was not an arrogant man. He was shy, tended not to talk up his own work, and was reluctant to accept enthusiastic praise from colleagues. No one had it in for Hendrik. He didn't have enemies. It's probably one of the reasons it went so far.

If he had been confrontational, someone would have gotten mad and had the incentive to go after him. He also wasn't in it for the money. He drove a mid-range car, rarely dressed up, and didn't go out to fancy restaurants or anything like that.

At one point he had several paychecks piling up at his desk in Murray Hill. And when Christian Kloc offered to deposit them, stating that one was missing, Hendrik didn't seem concerned. The one concern he did seem to have was keeping his full-time position at Bell Labs.

And with the uncertain financial state of the company, his productivity would likely be the deciding factor. During a bachelor's degree, there tends to be an emphasis on getting the right answer, so you can wrap your head around the fundamentals.

As you advance in years though moving on through your Masters and later your Ph.D, the focus has shifted from getting the result you expected to explaining why you got the less convenient result that you did see.

Somewhere during his Ph.D, this just didn't click for Hendrik. Hendrik's overly literal interpretation of physics was sending him down a dangerous path. Even from his time at the Universität Konstanz he tended to alter or muddle his original data in order to make the presentation of his results clearer. Hendrik developed an aversion to outliers, to noise.

It's as if he viewed them as something to be eliminated from his data and not a fundamental reality of his experiments. He would replace data with best fit lines, he would smooth data out or he would connect data from two different samples as if they were from the same sample.

Hendrik was fabricating nearly all of it. He wasn't plagiarizing, he wasn't cherry-picking, he was making it all up. Because he fundamentally did not grasp the scientific method. Hendrik learned early on that not all his work was going to be publishable. That's part of the brutal reality of academia, but if he couldn't put something out soon it might cost him his job, like it did hundreds of his co-workers.

So, he made one essential change to his scientific process.

He did it in reverse. He decided what results he wanted to see, and constructed the data backwards until he had a full set of graphs.

That's why he was so successful in every area he dipped his toes into. He knew what results would excite the scientific community, get them talking. He just needed to make that data seem plausible, and the way he did it was genius.

When people would come to Hendrik with questions or concerns, he would change his data to match their expectations.

Here's an example: Early in 2000 Hendrik was reporting that he observed superconductivity in molecules of carbon-60. He showed beautiful curves where the resistance plummeted to zero as the temperature approached 11 kelvin.

Harold Hwang, a colleague, asked Hendrik if he was sure that it was genuine superconductivity, or if his measurement setup might be seeing a short circuit, which is when an unexpected electrical connection occurs. Hendrik asked Harold what experiments he should perform that would convince him it was indeed superconductivity.

Harold gave him some follow-ups he could try, and Hendrik came back soon after claiming he did one, even though he hadn't. Hendrik was more than happy to accept feedback because it meant he had new ammunition to make his fraud more convincing.

This tendency to agree with everything made it impossible for others to track their own progress.

At one point, Kloc switched his crystal setup to use Hydrogen gas instead of Argon, hoping it would improve the growth conditions for his crystals. When Hendrik reported back to him that he saw an increased conductivity, Kloc assumed his switch had led to the improvement. It was only later that he realized Hendrik was just artificially reporting better results because he knew that's what Kloc expected to see.

Peter Littlewood, who had previously spent summers at Bell Labs, described the situation this way:

"You would say something and then it would

happen. You would get caught up in the progress of the subject. For a long time, I didn't believe it could have been fraud because I didn't believe one person could make it all up. Then I realized we all made it up."

Littlewood later said that Hendrik Schön was the "best listener of any experimental physicist in the world."

Hendrik would publish a new result perhaps with a bizarre, unexpected trend. Someone would suggest an explanation and then miraculously, the next few papers Hendrik produced would align suspiciously well with that proposed theory. Although this worked for a time, people gradually caught on to this small tactic of Hendrik.

The problem, Forrest said, was that Hendrik Schön seemed to be expecting the rest of the scientific community to do his heavy lifting for him, reproducing work that he hadn't done very carefully to start with, and failing to provide a clear interpretation of his results.

Over the years, rumors have circulated there were in fact negative reviews of Hendrik's submissions that were ignored by journals because they were more concerned of publishing exciting attention-grabbing headlines.

This is nearly impossible to confirm because reviewers are always kept anonymous.

Nature has stated that in general they were willing to discuss the specifics of the review process after a paper is published but they declined to do so in the wake of the Schön scandal because they, and I quote, "lost many of the records related to their handling of Hendrik's manuscripts."[9]

Science, on the other hand, has publicly stated that they will blacklist any reviewer who speaks publicly about papers they've reviewed.[10]

Eugenie Reich, in her 2009 book on the scandal, did extensive research and ended up obtaining info regarding reviews for 17 of Hendrik's papers and none of them were blatantly negative. What she did find was many positive, but also conditional, acceptances. And Hendrik used his natural agreeable personality to exploit the hell out of these conditional acceptances.

Many reviewers, despite being impressed, had specific concerns or suggestions. In principle, these reviewers were saying: If you make these changes or clarify these things, I think this is publishable.

Hendrik loved using relative units on his graphs rather than absolute units.

So, for example, let's say you have a current versus voltage graph. In some cases, it's more useful to just put the current as a percentage of the max value, rather than a specific number of amps. That's because it's the shape of the curves and not the exact numbers that is of interest.

Now, this is fine for one or two graphs, but at some point, you do need to include exact numbers. The issue for Hendrik is that exact numbers are way harder for him to fake and would just expose him to more specific questions about his work.

The more specific a question, the harder it is to bluff. For one of his papers, a reviewer made this exact complaint. He gave an overall positive review but asked for the inclusion of exact numbers. Hendrik received this critique went ahead and included exact values in amps and the next time that person saw the paper it was already in print.

To his surprise, he found that the current Schön reported having measured across the device was unrealistically small. This should have been impossible to measure.

Suddenly a paper that he had recommended enthusiastically for publication was beginning to unravel. Hendrik was always careful to avoid responding aggressively. He would write back in a polite, formal manner, asking for advice and suggestions from the reviewers, acknowledging their critiques as valid.

But when he resubmitted his manuscripts to the journals, he almost never made any meaningful changes. When in doubt, Hendrik's most popular tactic was to cheerfully include: "further experiments will be needed" and he was off the hook. By acknowledging his work left many unanswered questions he managed to skirt by the review process.

In one particularly sneaky case, Hendrik submitted the paper on his molecular transistor to Nature. One reviewer found his claim of molecular scale computing to be exaggerated, and that it should be deleted from the final submission.

Hendrik got around this by saying the necessary data to make that claim was going to be in a soon-to-be-published paper in Science.

He was playing two competing journals off each other. This lack of specifics regarding his experimental setup also made it harder for other researchers to try and replicate his work.

If just one other lab could reproduce Hendrik's work, it would dispel any doubt that Hendrik was taking home a Nobel prize. But at that point nobody else had sort of duplicated anything yet.

But by 2002, it had been over two full years since Hendrik's breakout paper on high conductivity pentacene, and not a single research group had been able to get anywhere close to his conductivity values. In physics, two years is an eternity.

What was one man doing differently that dozens of other labs couldn't?

9 SPUTTERING OUT OF CONTROL

It's worth a reminder that throughout this whole story, Hendrik remained a German citizen, which means that to legally work in the United States, he needed a visa. He got one for his four-month internship, then he went back to Germany for four months to finish his Ph.D.

To return to Bell Labs, though, he had to reapply for a new visa, since the terms of his employment would have been updated. He was no longer an intern but a postdoc on a two-year contract. As we learned earlier, no postdoc at Bell Labs is guaranteed a full-time job, especially given the financial circumstances.

But Hendrik was the rare exception to the rule and in December of 2000, he officially got his full-time contract. But again, because the terms of his employment had changed, he was obligated to return to Germany and apply for another visa, a process that lasted five months.

During this time, he still worked for Bell Labs. He

just worked remotely in his old lab at the Universität Konstanz.

Here's where it all starts to come into focus: This five-month period in Germany is responsible for almost all his breakthroughs in 2001.

Hendrik received no shortage of praise from his managers at Bell Labs, but given the economic state of the company, chief among them was the fact that he was cheap. He rarely, if ever, made funding requests to his department directors and yet he still managed to make several breakthroughs. This became so prevalent that other employees were even chastised in performance reviews and told that they should work more cost effectively.

So, the fact that he was:

A. cheap,
B. in a separate country, and
C. a breakout scientific rock star,

meant that Hendrik had little to no oversight and was largely left to his own devices. Besides the organic crystals provided to Hendrik by Christian Kloc, he didn't need much. The Universität Konstanz had every piece of equipment he could ask for.

Most of his work was slapping metal electrodes on top of crystals and measuring the current going through them. All of that could be done at a simple lab bench. This sort of thing was stupid simple, which meant that nearly any rival research lab could be doing the exact same tests at little cost.

As the story goes, anytime Hendrik was asked what his secret ingredient was he would always give the exact same answer: Aluminum oxide. General rule of thumb:

When making transistors, the more electrons you've got flowing, the more conductive it is. So, you'll have millions, maybe even billions of electrons flowing back and forth in a confined channel. And since Hendrik was also making superconductors, he was squeezing absurd amounts of electrons into the same space.

Careful though, that's a lot of pent-up volts that need somewhere to go. The critical element here is the top insulating layer. You want the electrons flowing this way, but they really want to burst out this way.

You need a material that can act as a shield that can withstand that buildup of charge. If it's too weak, the electrons punch a hole through it, maybe even setting fire to your chip.

Typical transistors use $SiO2$ as their insulator, but the charge densities Hendrik was reporting needed something stronger. Which is why he used aluminum oxide. This fact on its own wasn't innovative on Hendrik's part. The remarkable thing about his results was that, even with a stronger insulator, he claimed to observe charge densities in the range of 70 to 80 million volts per centimeter.

That's a lot of volts. These fields were so strong they'd annihilate a regular transistor and is about five times higher than what every other lab said they could handle. As long as no one else could achieve the same charge densities, he had plausible deniability he was doing something better than everyone else.

Here's where the big miscommunication came into play. When researchers asked Batlogg how they could make high quality aluminum oxide, Batlogg implied that they would need to invest in high quality sputtering equipment.

This led people to believe that Bell Labs had a high-quality sputtering facility, when in reality, Hendrik was traveling back and forth between Bell Labs and his old Ph.D lab at the Universität

Konstanz to use the fairly old and definitely not state-of-the-art sputtering machine there.

All of Hendrik's key breakthroughs can be traced back to one machine, that wasn't in his place of work, and required a 10-hour flight to get to.

Very interesting.

By this point, Hendrik was a bit of a celebrity, and often received emails from labs around the country asking for advice on replicating his experiments. Colleagues continually pestered him about how he was sputtering the aluminum oxide, since this was the experimental step that nobody seemed to be able to replicate.

Hendrik must have realized that people would eventually catch on because he started rewriting history to lend himself credibility. Rumors circulated that Hendrik had mastered sputtering machines when he used them for his Ph.D thesis, which wasn't true. He didn't use them for his Ph.D thesis, but Hendrik would support those rumors when asked.

The logic at the time was that if there were any imperfections in the aluminum oxide, it would disintegrate at high voltages, so it needed to be really, really pure. But what didn't make sense was how

Hendrik, who was working on some underutilized piece of equipment in the corner of an obscure German laboratory, was somehow producing better quality oxide than labs with much better equipment.

The only other explanation was that there was something specific about the Konstanz machine like a unique combination of atmosphere and other chemicals that just happened to make a perfect oxide deposit. It was just too absurd to be believable.

It got so bad that at one point, colleagues were begging him to publish an entire paper just on the aluminum oxide process. He finally relented, tediously compiling results at dozens of deposition rates and pressures, in an unpublished document he circulated around to those who asked. Although he did submit it, the paper never officially got published by a journal.

But much like a lot of things in this story, it would still come back to haunt him.

[RETRACTED]

10 THE F-WORD

It's impossible to know for sure, but it's thought that the first recorded suspicion of fraud came over a year before the scandal went public.

It was in May 2001, in an email from Bob Laughlin, a theoretical physicist. He perhaps felt bold enough to make this claim because, well, he had worked for Bell Labs many years ago and had no problem criticizing his former place of work.

But if I had to guess, it was mainly because he had a Nobel prize under his belt.

Bob Laughlin had serious doubts about the deposition of the aluminum oxide. His knowledge of the physics of insulators didn't match up with what he'd heard from Batlogg. Laughlin even publicly criticized Batlogg in an August 2000 seminar, claiming that the data he was showing was a far too narrower range to be meaningful.

In the summer of 2001, he went as far as to approach Takehiko Ishiguru with his concerns, who then went on to publish one of the first critiques of

Hendrik's work. Bob Laughlin had a reputation for being outspoken and highly skeptical in general so his word alone would not have been enough to spark widespread suspicion.

Doubt? Sure, lots of people were confused about Hendrik's exact methods. But fraud? Not quite yet.

So why not?

Why was there such an aversion to speaking up when so many people either couldn't reproduce or didn't even believe Hendrik's results? Because of the taboo. Science is a difficult line of work, and it only gets done because of collaboration.

Jumping to accusations right out of the gate when there might be a much more innocent explanation is never advisable, not to mention the potential to get it wrong. Accusing an innocent person is risky and might turn into a public scandal that can drag on for years. This is made even harder the more famous the accused is.

You wouldn't just be accusing Hendrik. By extension you'd be accusing Batlogg, you'd be accusing the entire institution of Bell Labs. Without a smoking gun, indisputable evidence of fraud, it's an uphill battle where it's your credibility against an institution, a scientific dynasty.

If you happen to be wrong, you deal a permanent blow to your own career. If you happen to be right, well, you did the right thing. But every new case of fraud erodes public trust in science and makes it harder to get funding and resources.

Not to mention it can lead to fearmongering about vaccines and 5G cell towers. And even if fraud is a possibility, it's never the first thing people jump to. Maybe he made a mistake, maybe he just misinterpreted. Maybe I'm not as good at science as I thought I was.

As crazy as it sounds, Hendrik evaded discovery for so long because people kept giving him the benefit of the doubt. Getting a paper out of Bell Labs was supposed to be more difficult than a university.

"In academia I can write up and submit a manuscript to a journal without having to show it to anyone or getting anyone's approval. At Bell Labs, a draft paper was then sent to somebody else in the division, not directly connected with the work, for a once over before it could be submitted. The assumption was that these checks were being performed on Hendrik's work."[11]

In truth, this wasn't happening. Most of the time, Hendrik would run his papers by Batlogg and no one else, sometimes ignoring the many experts in the building who would have been able to provide invaluable commentary.

In one case a co-worker, Art Ramirez, approached Hendrik with an organic crystal called perylene. He expressed interest in measuring its electrical properties, which was dangerous for Hendrik because it meant someone would be double checking his results. Hendrik tried to dissuade Ramirez by claiming he had already worked with perylene before.

But Ramirez insisted. The next week, Hendrik presented Ramirez with a freshly typed up manuscript that reused old graphs, but with perylene swapped in for the figure title. This finally convinced Ramirez to back off. Hendrik was willing to go as far as to convince people he had articles ready to publish just to get people off his tracks.

When another co-worker, David Muller, asked Hendrik if he could inspect his laser under a microscope, Hendrik said he'd bring it back from Konstanz on his next visit. When Hendrik later did no such thing, he told Muller he forgot it.

Lucky for Hendrik, however, Muller interpreted

this as Hendrik's polite way of saying, 'This is my work, back off.'

In an environment like Bell Labs, especially at the height of its economic troubles, this wasn't suspicious. It would make sense for Hendrik to be territorial about a huge discovery like this. Benefit of the doubt helped Hendrik in other ways.

Joe Orenstein of Berkeley had his group attempt to analyze the organic crystals that Hendrik had used in his superconductivity experiments. This meant cooling the crystals close to absolute zero. Big problem though, the crystals always shattered at 100 kelvin. Nowhere close to the temperatures Hendrik was reporting.

This, of course, was because Hendrik hadn't performed the experiment and wouldn't know the crystals would shatter, but again, rather than assume there was a problem with the Bell Labs work, Orenstein assumed either there was something wrong with his experimental process or that the quality of his crystals was just not good enough.

Of course, he was assuming that Hendrik was operating in good faith, which you know he wasn't. This assumption often extended to how Hendrik interpreted his results as well.

Hendrik was not an expert at lasers, and neither was Batlogg. Before he even published what would become his second most cited paper, there was a rather heated internal debate over what his results actually meant. In his paper, Hendrik presented two intensity-versus-wavelength plots supposedly taken at two very different temperatures.

Both shapes are remarkably similar Gaussian curves, which is odd enough for such different temperatures, but you'd also expect to see the graph to shift horizontally too, and that clearly hasn't happened here.

In fact, Hendrik's laser spectrums were so baffling that no one could agree on whether Hendrik's crystal was lasing, or whether it was demonstrating another effect known as super luminescence.

The debate wasn't over whether Hendrik's laser data was real or not, but whether he had actually made a laser in the first place.

The molecular transistor was Hendrik's final claim to fame, and it was such a bold breakthrough to fake that it ultimately led to his downfall. For months, though, he managed to fend off several tough questions regarding his confusing data.

IBM is one of Bell lab's fiercest competitors. They hold the second-place record for the most Nobel prizes won by a private company. At one point, a collection of IBM researchers were dead set on submitting a paper to Nature that was a direct critique of the molecular transistor paper which came just short of accusing Hendrik of fraud.

Nature reached out to Hendrik and Bao for a correction.

In the end, Nature decided not to publish the IBM critique, because Hendrik claimed he had performed more experiments to confirm his results weren't just a fluke and he had an upcoming paper in another journal which would clarify some of the finer unknown details.

Now, where have we heard that before?

During a seminar he gave at Bell Labs, Hendrik was questioned about two molecular devices of very different sizes. Rafi Kleiman didn't believe that two devices of such different sizes could be demonstrating the same physical effect. But Hendrik jumped in to say, 'oh no sorry about that that's actually a typo, those devices are the same size.'

This seemed odd at the time to Kleiman, who later checked the original submitted paper and realized that they were in fact different sizes. He correctly deduced that Hendrik had lied to his face, but Kleiman assumed it was because Hendrik didn't want to escalate a public argument, having no reason to think that Hendrik had faked the data.

From this same seminar, Julia Hsu asked Hendrik about an image taken from a microscope.

Specifically, she wanted to know which microscope he had used. Hendrik replied that it was not in fact his picture, but he took it from literature. Julia was shocked to hear this, considering Hendrik hadn't referenced anybody in the presentation, but opted not to elevate her concerns since the seminar had been relatively informal. These exchanges are prime examples of why people tended not to engage with Hendrik at public events.

He either created a moving target altering numbers on the fly or changing his explanations in response to feedback. David Muller is quoted as saying "Wait for it to get into print before you criticize it."

This fella, Don Monroe, didn't follow that advice. He was the first person to leverage an official accusation of fraud against Hendrik Schön. Don

Monroe was a scientist at the spin-off company Agere Systems, which still technically fell under the Bell Labs umbrella, so employees of the two companies often discussed their work as they ate lunch together.

Since his interests were in ultra-small transistors, Monroe proposed an inter-company seminar to Hendrik, hoping to provide some feedback on his molecular transistor or at least a lively debate.

Hendrik agreed and as part of the seminar, he showed off the following histogram:

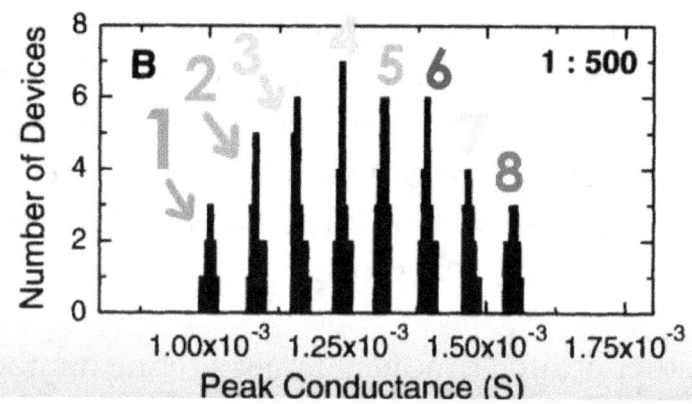

Here you can see he's quantified the conductivity according to how many molecules were in the device.

So, one bin is for devices with one molecule, the next bin is for devices with two, three and so on. When fabricating many copies of the same device you'll inevitably run into some variation in sizes, and since we're talking about the scale of individual molecules, you'd expect to see clumps of individual molecules just like this.

What's weird here is that you can see a pretty clear Gaussian distribution, A.K.A. the Bell curve. And that just makes no sense, look at the sample size. If you count all the bins on this histogram, you get a total of 130 devices.

That's nowhere near enough of a sample size to get a Gaussian this perfect. If you run the numbers, there's only a 1.2 percent chance he gets a fit this clean.

But it gets worse.

According to Hendrik, these are just the functioning devices, and those only have a yield of 10 percent. So, assuming you take Hendrik at his word, and he manually tested every single device that he made, that's 1,300 devices.

When's the last time you did 1,300 of anything? If we assume (quite generously) that he took five minutes to create and measure each device, that would take 108 hours, or four and a half days of non-stop work.

This statistical anomaly didn't go unnoticed by Don Monroe, who raised his concerns to Hendrik, but was quickly brushed off. Ten days later, this histogram appeared in print. Unsatisfied with Hendrik's explanations, Don Monroe escalated the matter to Hendrik's managers.

He was cautious about his wording, never outright saying that fraud had been committed but his statistical analysis of Hendrik's data very heavily implied that it was the only explanation.

It turned out that even prior to this email, Hendrik's managers had been telling him to pump the brakes on his publishing output as concerns over his data and lack of experimental details had been rising. This resulted in at least one paper on a single electron transistor being held back from submission.

Over the coming months Hendrik was asked repeatedly if he had some explanation for the histogram ending up so symmetrical, which led to him eventually sharing a much noisier histogram that was far more believable. For his managers and Don Monroe, this was a satisfactory resolution to the matter. Again, assumption of good faith.

Nobody wanted to escalate things further than

they had to, especially if an explanation had been happily provided. Hendrik's manager John Rogers is quoted as saying, "while I am not averse to an investigation, any direct involvement on my part would destroy my professional relationship with Hendrik."

This would be the closest Hendrik would come to being found out and still getting away with it.

But in a few short months, it would all come crashing down.

11 WHISTLEBLOWERS

You remember Julia Hsu?

She and Lynn Loo were both working the same department as Hendrik in April of 2002. While at a conference, they were discussing a new process for fabricating circuits.

By April 19th, they had enough data for a patent application, and together they visited Bell Lab's patent attorney, where they began sifting through papers on similar concepts in order to write up their patent and justify why their technique was innovative. Two of the papers they were sifting through were on Hendrik's molecular transistor.

Lynn noticed that two of the graphs appeared oddly similar considering that one of the devices was for a transistor with thousands of molecular layers, and the other one for only one or two layers, but it went beyond that.

The noise in the graphs was identical, down to the pixel. Two experiments on the exact same device at the exact same temperature on the exact same day should not give the same results. Much less two completely different devices.

Both Julia and Lynn immediately told colleagues about the duplication.

Lynn went to her supervisor Bob Willett.

Now, Bob Willett, not exactly a fan of Hendrik.

For a long time now, internal tensions over the believability of Hendrik's work were beginning to boil over.

Two months earlier, Bob Willett had been awarded the $10,000 Oliver E. Buckley Condensed Matter Prize and a symposium was being organized in his name. Bob Willett was upset to find out Hendrik was scheduled to speak since he didn't have faith in his results, and ended up getting into a shouting match with Frederico Capasso, Hendrik's department head.

In the end Capasso won out, and Hendrik spoke at the symposium.

So yeah, Bob Willett, not a fan.

After Lynn came to him about the duplicate, he

immediately went to John Rogers, Hendrik's manager.

Julia, at the same time as this, went to Mark Lee, her colleague and husband, asking him if he could think of any reason the graphs were so similar. Perhaps an artifact of the measurement setup?

Mark Lee then coincidentally also went to Bob Willett to ask if anybody else knew about the duplication. Mark assumed that Hendrik had just made a mistake and copied in the wrong graph. Bob Willett, who already doubted Hendrik's results, assumed Hendrik had been duplicating data in several papers, as it was the only way to explain his rapid publication pace.

Lee approached Hendrik about the duplicate. Hendrik looked embarrassed and admitted that 'oh, looks like he made a mistake.' Lee then suggested that Hendrik send a correction to Science.

Lee relayed this encounter with Hendrik to Bob Willett, who then phoned one of his old postdocs Lydia Sohn, now a professor of physics at Princeton, telling her to keep her eyes open for an upcoming correction in a particular paper of Hendrik's.

Whether Bob Willett can technically be considered a whistleblower is up for debate. From his perspective, he was just informing an old student about something he thought was public knowledge, that Hendrik was issuing a correction to one of his papers.

But Bob couldn't have predicted what would happen next.

Lydia Sohn had been suspicious of Hendrik's papers for some time now. At one point, she had discussed the molecular transistor with a colleague who was adamant that the data was fake.

His evidence? The perfectly symmetrical histogram.

Little did they know that an internal investigation had already been brought up and resolved over this exact figure. Even based off nothing but the vague hint from Bob Willett, Lydia Sohn quickly figured out the problem with Hendrik's paper.

But it couldn't have just been a mistake, the curves in these graphs, they're identical.

Hendrik even changed several numbers and the caption, that's a lot of work to do accidentally.

Sohn quickly emailed multiple colleagues at different universities and unanimously agreed there was a serious issue here. They all recommended that Sohn contact Science and Nature, since she was the only one who wasn't in the process of submitting a paper to either journal.

At the same time as all of this, John Rogers, who Bob Willett had talked to directly, had been conducting his own investigation into the duplicate figures. He ended up uncovering some figures that looked similar, but not exact duplicates.

He just wouldn't have known to look back as far as two years earlier to a completely different organic transistor that ended up being the original source for the duplicates Sohn had identified.

For those keeping score, we're up to three exact duplicate curves for three completely different devices in three completely different papers across a span of two years. The discovery of the third duplicate is credited to Paul McEuen at Cornell.

McEuen was one of the first people Sohn informed about her discovery. He showed his findings to Dan Ralph, who is on the committee for the William L. McMillan Prize, which Hendrik had nearly won in 2001 but ultimately did not get because no one had been able to replicate his results yet. But a year later, he was the clear favorite to win for 2002.

McEuen's intervention is the reason Hendrik's prize letter was never sent.

McEuen compiled a refined set of the six most compelling pieces of duplication he could find and sent it to Hendrik, John Rogers, Batlogg and editors at Nature and Science.

On May 23rd, the story broke in the New York Times.

Too late to turn back now. Bell Labs' reputation was hanging by a thread, and it all depended on how they conducted their investigation.

One report, five committee members, 24 accusations, 127 pages.

A report that would revoke a Ph.D.

12 PROPERTY OF LUCENT TECHNOLOGIES

In May of 2002, Bell Labs formally created an investigative committee. Prior to this, however, Hendrik's managers Rogers and Capasso reached out to Batlogg.

They asked him to speak with Hendrik directly and ask the question nobody had thought to ask:

Did you fake the data?

Hendrik's answer: No.

The committee was made up of five members and included a mix of experts on the relevant topics of Hendrik's publications.

The chairman, Malcolm Beasley of Stanford, has an extensive history with superconductivity.

Purdue University's Supriyo Datta focused on nano electronics and had been following Hendrik's molecular transistor for a while now.

Next, you had Herbert Kroemer, co-winner of the Nobel Prize in Physics in 2000 for his work with high-speed semiconductors.

Then there was Bell Labs' own Herwig Kogelnik, the go-to for all things lasers.

The most interesting inclusion, in my eyes:
Don Monroe - the man who first raised the idea of an investigation over the overly symmetric histogram, and the closest anyone had come to outright accusing Hendrik of fraud.

It was hard to argue that Bell Labs was planning a cover-up.

In short, that meant that three out of five members were external leaders in their respective fields, and one internal one who was certainly not afraid of getting to the bottom of things.

The New York Times had gone public with the story on May 23rd.

Bell Labs' reputation was in question at one of the worst financial periods in its' history. Transparency wasn't a suggestion, it was mandatory.

After a five-month investigation, the committee's

findings were compiled in a 127-page report: Property of Lucent Technologies.

The management had several audiences for the report. One was their own - the company - the Lucent management. The big company.

Another was the outside community, and another was the members of their own scientific staff.

Some of the internal critics were painted with the idea that somehow this was sour grapes, that they were just jealous. But it's hard to know what was going on behind the scenes. The committee first sent out a questionnaire to Hendrik which included asking him for all of the relevant original data files.

In some cases, they managed to find the original data embedded in journal submissions, rough drafts, and even PowerPoint slides. They also sent out questionnaires to all 20 of Hendrik's co-authors who were involved in the papers being investigated.

Of these co-authors, three were considered primary due to the frequency that they appeared on Hendrik's work.

These three were Bertram Batlogg, Christian Kloc, and Zhenan Bao. Batlogg and Kloc made sense to single out, they were co-authors on over 70 papers together. Bao was less obvious, since they only shared 14 papers.

However, those 14 included Hendrik's molecular transistor which was the target of multiple accusations. It became clear from their questionnaire answers that none of the three of them had ever seen Hendrik measure even a single one of his devices.

Batlogg was just Hendrik's supervisor, guiding and discussing, but never personally seeing the experiments himself.

Christian Kloc's involvement was limited to growing the large organic crystals, which he then handed over to Hendrik and never saw again.

Zhenan Bao had a similar situation. She provided Hendrik with the organic molecules he used in his molecular transistor. Although she produced a single graph in one of the suspicious papers, it was clear that there was nothing unexpected about her data, and that Hendrik had acted alone.

For over a month the committee was inundated with accusations, and at one point had to declare that their report would only consider accusations brought forward before June 2nd, otherwise the process would just go on for months.

After considering them all, they dismissed certain accusations as insubstantial and categorized all the remaining ones into three groups:

Data substitution - nine counts.
Unrealistic precision - nine counts.
Contradictory physics - six counts.

The full report dives deep into each accusation, but for the sake of time, I'm only going to be highlighting the most unique accusations.

For nearly three years, Hendrik's data was conjured from a black box.

The committee's job was to uncover what was inside.

13 INTO THE VOID

Of the three types of accusations, data substitution or duplication was the most blatant, and the easiest to prove.

Some of these you could find via visual inspection just as Lynn Loo and Julia Hsu had stumbled upon during a late night in the patent office. You've all seen accusation three by now. This is what kicked off the whole ordeal:

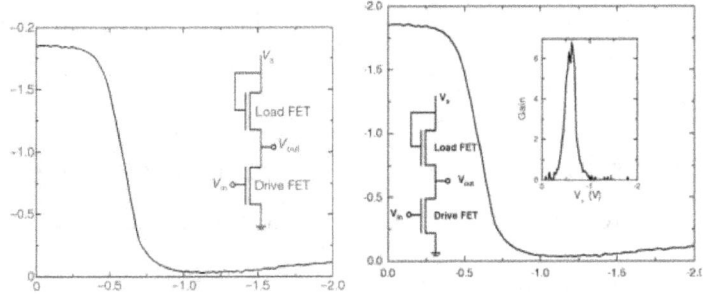

Three identical curves including noise, in three different papers for three different materials across a span of two years.

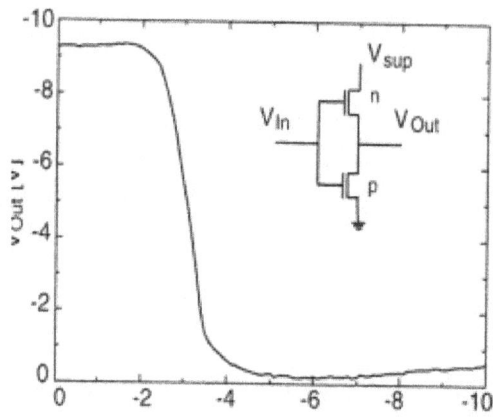

You'll notice the third one has a different numerical scale than the other two but the exact same shape, almost as if he was drawing it with a pencil. Cut and dry case, clearly misconduct on Hendrik's part.

Data substitution isn't always about one-to-one duplicates though, some cases were a bit more subtle.

Take accusation five: these two bottom curves look unrelated... right?

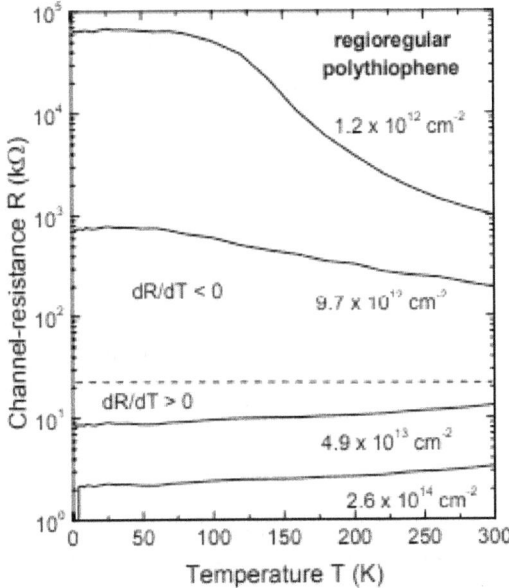

Turns out if you look at the raw data, they're exactly off by a constant factor of 3.96.

Hendrik made one curve and multiplied that by what the theory suggested would be an appropriate factor. He clearly realized that multiplying by exactly four would be suspicious, so he adjusted it to look a bit more plausible. That's not even the worst one, though.

Accusation one highlights two different figures from the same paper of what is apparently two different molecules.

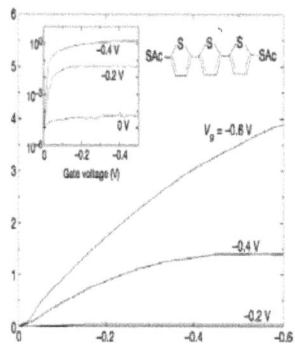 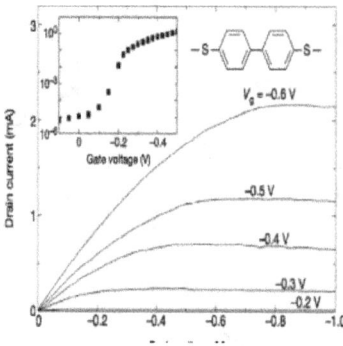

The curves differ exactly by a factor of two. Not 2.01. Not 2.001. I'm talking 2.0000.

Contradictory physics, I'd argue, are the most fascinating accusations. That's because they're by far the most difficult to prove. The idea is that Hendrik's results contradict some generally well-accepted piece of theory, and no plausible alternative explanation can be found.

The trouble is Hendrik's most famous publications covered exotic materials no one had really worked with before, making lasers and superconductors with materials no one else had. There was a tiny, but still real chance that some of the contradictory physics were a legitimate discovery on Hendrik's part.

After all, many key pillars of physics like relativity and quantum mechanics got their start as contradictory physics. These accusations are partly why such high-profile researchers were chosen for the committee. If anyone could figure out what was real and what was fake, these five had a good shot.

Accusation 19: One of the most basic building blocks of circuits is the inverter. If it takes an input of one it outputs a zero, if it takes an input of a zero it outputs a one, simple!

Hendrik's voltage curve looked like this, the top arc with a nice, distinct separation between the high and low regions.

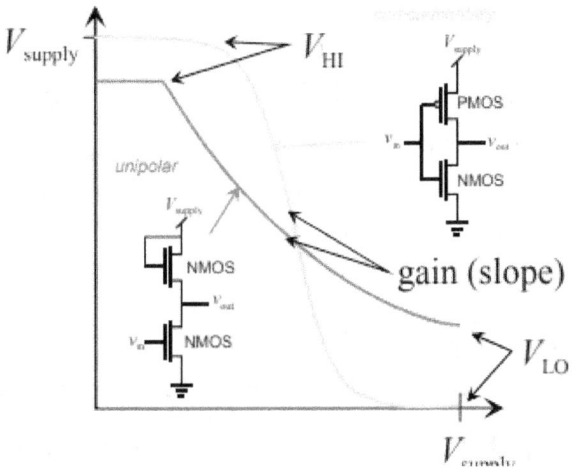

This is the iconic inverter curve associated with two transistors hooked up like this. The issue here is that the inverter circuit that Hendrik describes in his paper actually has one transistor and a resistor, and that type of inverter produces a very different curve, seen on the bottom.

It has a very sharp, flat plateau and a much more gradual descent. This contradicted common knowledge for one of the most important circuits ever invented.

Similarly for accusation 20. Hendrik's molecular transistor design - the first of its kind - had a pretty unoptimized geometry, performing far better than even the cutting-edge transistors that had decades of research behind them.

So, you know, not really believable.

Compared to data substitution, accusations of unrealistic precision are much less visually obvious, but because the committee eventually got access to the raw data files it gave them the chance to run some rigorous analysis on his numbers.

In several instances Hendrik would have graphs where the scales just seemed too large to be measurable, or the curves just looked way too good to be true.

You'd have a pretty standard curve for a superconductor: as the temperature approaches absolute zero it abruptly drops off and then goes to zero.

Except the superconductor in question shouldn't have such an abrupt transition. It's generally expected that the transition point has a lot more fluctuations. So that alone was already suspicious, but looking at the raw data, nearly all fit a mathematical expression or are exactly zero.

That leaves only four points out of a total 117 that might... just might... be genuine data.

Finally, accusation 18. This one is a big one. Hendrik's unpublished paper on his aluminum oxide sputtering. This was supposed to answer all the questions of researchers who couldn't recreate his results.

Let's see how it holds up.

Hendrik claimed, in multiple papers, that his aluminum oxide could support up to 70 to 80 million volts per centimeter. The most any other group could achieve was 12 to 15 million, a factor five times lower.

Hendrik sampled 600 devices and measured their breakdown field, which is the field where the devices fail and get torn apart. It resulted in a gaussian distribution like you see here,

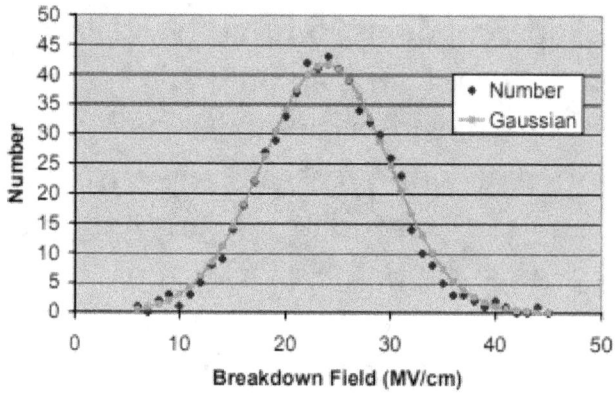

except 600 points is far too low a sample size for such a good agreement and the chance of that occurring is 0.02 percent... so not super plausible.

And as it turns out, a Gaussian isn't even the expected distribution you'd see here. You'd anticipate seeing a Weibull distribution, which, for someone doing a half-assed fake, might confuse with a gaussian at a glance.

The next figure in question is a contour map of pressure versus deposition rate with red spots indicating the best results. Hendrik claimed he recorded 144 different deposition rates, and he reported his breakdown field with enough decimal places that it would have required at least 36 measurements to be accurate.

What that means is he would have had to take a total of 36 times 144 total data points which works out to the fabrication and testing of 5,184 separate devices!

And you thought 1,300 was bad.

If we assume, again quite generously, that it takes him five minutes to both make and test each device it would have taken Hendrik 18 straight days of non-stop work. Hendrik didn't actually measure 5,000 data points.

When asked, he claimed he expanded his 36 data point set to over 5,000 by interpolating between points. Which is a completely meaningless and information destroying procedure for an experiment explicitly meant to show process variations.

Finally, it was discovered through an email conversation with Batlogg that he initially claimed he was going to sweep his voltage at a rate of 0.001 volts per second. Batlogg pointed out that it would take 2.4 years for that to finish, and he promptly changed the voltage sweep from a value of 0.001 to 1 volt per second... a factor of 1000 difference.

The resulting plot from this voltage sweep was the following.

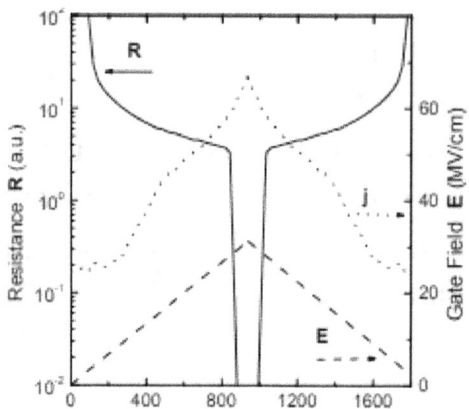

Note how the data is perfectly mirrored at the point where he starts to sweep in the other direction. And here we are. At the end of the day, the most blatant and error-filled paper of Hendrik's was the one he never wanted to publish in the first place.

His secret ingredient had always been his aluminum oxide, and when put under a microscope his excuses and explanations fell apart.

And if he was lying about this, every paper he had ever touched was in doubt.

14 [RETRACTED]

Hendrik was found to have committed 16 out of the 24 formal allegations of scientific misconduct.

6 out of 9 for data substitution, 7 out of 9 for unrealistic precision, and 3 out of 6 for contradictory physics.

Note the carefully chosen language on the ones he wasn't found guilty of:

"while troubling, this instance on its own does not provide compelling evidence of scientific misconduct" or "the preponderance of evidence indicates that Hendrik Schön fabricated data in this case, however since no relationship between the fabricated data and any published data has been established, no finding of scientific misconduct is warranted."[12]

The data they were able to gather came from submissions Hendrik made to journals where the data points were still embedded in word documents, or graphs copy pasted into PowerPoint slides. He claimed it was standard practice not to keep written logs, which is not true.

What about digital copies?
Still none.

What do you think the official reason he gave for that was?
I genuinely want you to guess.

I want you to make up an excuse so bad you'd never even dream of using it because you'd be too embarrassed to pretend it was even real.

He deleted them because he ran out of hard drive space.

So, you're telling me that a man famous for breakthroughs in several disciplines, who was known for doing some of the cheapest research at the company, didn't think he could ask for an extra hard drive?

It shouldn't come as a shock that during the investigation they found that Hendrik had not a single working, or even non-working, device for them to analyze.

He claimed that all of them were:

A. Damaged during measurement, which, sure, let's assume high voltage breakdown led to some occasional broken devices, that's plausible.

B. Damaged during transit from Konstanz, uh all right sure I'm playing along, I'll humor you.

C. *He threw them in the garbage.*

…

Hendrik was fired the same day the report was published.

Hendrik maintains that even though he falsified the data in his publications, he was observing a genuine physical effect, and that further research will validate his findings. Whether that happens or not, it sounds like he fundamentally doesn't understand what he did wrong.

All those awards he won? After weeks of negative headlines, he returned the prize money, alongside his co-winners.

The Schön scandal remains one of the most prolific and highly publicized cases of academic fraud in living memory.

In 2007, Nature even published a misconduct special with Hendrik front and center.

Among all these cases, the Schön scandal is unique in that it's one of the few fraud scandals where the central figure was unequivocally found to be guilty, and the law allowed for them to be stripped of their academic standing.

After a lengthy investigation of their own, the Universität Konstanz revoked his Ph.D in 2004. This fact is a lot more significant than you might initially think, as revoking a degree is extremely rare, as well as a legally complex process, and it often comes down to the provincial or state laws where the university is located.

Even then, the authority to do so is usually reserved for cases where there is clear misconduct associated with the Ph.D thesis itself, most often plagiarism.

The university conducted an investigation of Hendrik's thesis and noted that although they found mistakes, they never actually found any clear evidence of fraud, but still deemed Hendrik unworthy of his doctorate.

For this reason, Hendrik sued on the grounds that Konstanz didn't have the authority to revoke his degree, and in 2009 the administrative court of

Freiburg agreed with him.

However, in 2011, the administrative court of Manheim argued that under the State Higher Education Act, unworthiness can relate to scientific conduct.

To paraphrase the legal decision: A Ph.D is an endorsement of an individual's research abilities, and the title holder enjoys an increased level of trust in the eyes of both the scientific community and the public.

Therefore, a serious violation of scientific ethics warrants the removal of the degree. This decision was upheld by the federal administrative court and the federal constitutional court declined to even view Hendrik's additional complaint.

As of 2011, Hendrik is no longer a doctor.

On top of that, the German research council banned Hendrik from receiving research funding for eight years, as well as barring him from acting as a peer reviewer for journals.

He's still allowed to work in physics, mind you, just not as a doctor. He now works as a process engineer at a firm in Germany. So, why was Hendrik never criminally charged?

Well, the US federal policy on research conduct only applies to research that is government funded. Because all research was internally funded by Lucent, a private corporation, no criminal charges could be applied.

The most lasting consequence though, was on its publication record.

The Bell Labs report led to 28 papers being retracted. On top of that, eight others were deemed highly questionable, which is a polite way to say we wouldn't trust anything in this.

This works out to almost a third of Hendrik's entire career output being deemed invalid.

And despite all that, Hendrik is still only the 10th most retracted author of all time.

15 EXTRAORDINARILY DIFFICULT QUESTIONS

It's hard to argue that Hendrik didn't get exactly what he deserved. Far better people have lost their jobs for doing far less than what Hendrik did.

The much more hotly debated question is whether his colleagues shared proper responsibility. Although the committee's report goes out of its way to state that only Hendrik was guilty of misconduct, there is a separate question of whether his co-authors met their professional responsibilities.

Co-authors, by their very nature, are implicitly endorsing any piece of work their name appears on. Collaborative scientific research requires a high level of trust among the participants; however, such trust must be balanced with a responsibility to ensure the veracity of all results. Shared credit for the accomplishment must be matched with shared responsibility.

But the committee were also aware that someone like Christian Kloc, who was providing samples, would not have the expertise to evaluate the electrical measurements, so holding everybody responsible for the whole paper doesn't seem like a constructive way to deal with it.

Operating under this standard, the committee determined that all of Hendrik's co-authors met the minimum responsibilities, but did not go as far as to say they were entirely blameless.

In particular, they questioned whether Batlogg, as the distinguished leader of the research, took a sufficiently critical stance. The argument being that exceptional results demand exceptional skepticism, and Hendrik's work certainly was exceptional.

Batlogg, more than anyone, was in a position to cross that line of trust and question the integrity of Hendrik's results. Maybe you'd like to think that people judge papers solely based on their content, but reputation is a very important part of science.

People work very hard to get a reputation, and when someone has a reputation, they get an automatic degree of credibility. If there had been a Nobel prize, it seems likely that Bertram would be on the Nobel prize as well.

Despite their critiques the committee stopped short of calling for further disciplinary action for anyone other than Hendrik, "These are extraordinarily difficult questions which go to the heart of what we as a community of scientists expect of one another professionally. The committee does not consider itself qualified to make a specific judgment in this case in the absence of a broader consensus on the nature of the responsibilities of participants in collaborative research endeavors."

What about the journals? How accountable are they?

Speaking in general terms, editors at both Science and Nature emphasize that while they try to run a thorough review process, they do not aim to guarantee that everything they publish is right. Our goal is the advancement of science, does it advance science?

Even a paper that's wrong can encourage new science. Even if that is the case, should there not be a re-evaluation of the checks and balances in place, make it so fraud can't slip by undetected?

No, answers Donald Levy.

"Referees need to assume that the authors are operating in good faith. Once you admit the possibility of fraud, it becomes nearly impossible to evaluate the paper. It would be very damaging to the progress of science that every paper had to be screened for fraud."

This sentiment is echoed by Hendrik's former department head, Frederico Capasso.

"Scientists are good at detecting genuine technical errors but not at detecting fraud because their system is based on trust."

Yes, we were all fooled. Most people in in research don't think acceptance in a peer-reviewed journal is a mark of truth. It's not that simple, and there can be honest mistakes, but there can also be devious or intentional distortions.

As of writing, it has been over 20 years since the start of Hendrik's forgery. Nearly two decades of scientific progress. Was there any truth to Hendrik's work? Was he right, that he would ultimately be validated with further research?

The first molecular transistor was created, for real this time, in 2009 by a joint team at Yale and the Gwangju Institute in South Korea. He was only off by eight years.

So, how about Moore's law? I don't mean to be a downer, but Moore's law has been dead for more than a decade, and molecular transistors were never going to save it.

Moore's law was never a physical law. it was an economic one. The top chip manufacturers have been continually pushing back their smaller chip releases or canceling them entirely. It's not too difficult to lay down a single molecule connecting two gold wires.

The issue is connecting it to electrical contacts large enough to be useful, without noise dominating the signal from your electrode. Apply a slightly too large a voltage, you might physically be tearing gold atoms off the contacts. So, is that it? The age of digital progress has come to an end?

No. Computing power will keep growing, but the solution isn't going to be in the form of fitting more transistors onto a chip. The resurrection of Moore's law, if we even call it that, will come in the form of running several chips in parallel, will come from advances in solid-state drives, AI, machine learning, and maybe, just maybe, quantum computing.

Let's give that last one 50 years or so. Sadly, the plastic revolution he kicked off never really came to fruition. That's not to say organics is a dead-end field though. They have their own little niche. They're incredibly cheap to produce for large surface areas, can be made at room temperature, and have applications in flexible and biodegradable devices.

Plus, organic LEDs have dominated high-def TV displays for the past decade. Despite everything, the field of organics managed to survive, and arguably thrive, without him.

16 COLLATERAL DAMAGE

Of course, it's easy to look at the field as a whole and see little to no lasting damage, but on an individual level some were luckier than others.

Hendrik's paper kicked off a revolution. From the perspective of the academic community, organics was the new hot field and hundreds of eager grad students and postdocs bet everything on it. Many of them wasted months or even years trying to replicate results that weren't even real in the first place.

Funding applications scholarships and Ph.D proposals were unknowingly based on a foundation of lies. Bhattacharya surfaced to notice that his contemporaries were getting papers published and were being offered jobs as junior professors at good universities, and that he wasn't. His postdoc research, expected to be a short career stage lasting two or three years, went on to last five.

At one advisory meeting to assess the progress of Ruth DeBaur's Ph.D, he recalled the student practically in tears as she presented her mobility measurements and described Jan Hendrik's as better. DeBaur said that in this period she was dejected and thinking about giving up on her Ph.D topic which would lose time.

Goldhaber-Gordon's most experienced postdoc and a graduate student worked on organic crystals for nearly a year. When the fraud came to light, his grad students still had enough years left to switch to another project and graduate with a good dissertation, but his postdoc Silvia Luscher ended up having spent a year in her postdoc position without results, and after leaving the lab became less active in experimental physics.

Although she had other reasons for this, the year of frustration had not helped.

"You can't bring the time back. It was two years I could not use for my CV. It was not a good story to tell. It was my first real research experience."

Jochen Ulrich flew to the United States, visited Bell Labs, had an interview, and got a job, but to his disappointment he never got to explore quantum effects and organic crystals and had to spend months on the mundane task of trying to remake Schön's field effect transistors. Sick of the materials he switched to a new topic but still left science at the end of his postdoc.

"It was frustrating," he said. "I wouldn't go as far as to say that I left science because of this, but it changed the way I thought about science."

This is not even mentioning co-workers at Bell Labs who almost certainly lost their jobs when it might otherwise have been Hendrik. Some would argue that science is self-correcting. If someone is committing fraud, further research will naturally uncover them to be liars.

But as we saw in this case, Hendrik was not caught through passive research, letting things play out, but rather through active skepticism and aggressive investigation.

The Schön scandal is not just an embarrassing ordeal for a legendary institution, it's a stain on the field that set off a ripple effect of difficult personal and professional decisions.

Oddly enough, it seems that those most closely linked to the scandal went on to have largely unaffected careers.

Fredrico Capasso, arguably the most famous

scientist ever to work with Hendrik, took on a faculty position at Harvard where he continued to rack up awards and honors for his contributions to laser science.

John Rogers went to U Illinois and later Northwestern, working on flexible wearable electronics with 80 patents and rising.

Christian Kloc went on to have a productive, if modest, career as a professor in Singapore.

Zhenan Bao took a clear productivity hit for the first three years following the scandal, but afterwards she had an amazing output as a professor at Stanford. If anything, you could argue her career just had a delayed start to becoming one of the most productive chemists active today.

As you might have gathered, none of them remained at Bell Labs... which brings the question:

What happened to Bell Labs?

I wish there was a less depressing answer. Nothing to do with the Schön scandal, mind you. In the wake of the scandal, widespread policies were instituted that required all changes to remain struck out as part of the digital archive, and all pre-print manuscripts were to be hosted on an internal server for technical comments. There's a debate to be had over whether the leadership admitted sufficient fault over the scandal, but truth be told, the corporate end suffered very little from the fallout.

The changes the company was about to undergo were part of a much larger inevitable picture. As it's done several times throughout its existence, Bell Labs transferred ownership. Lucent, an American company, merged with Alcatel, a French company, and in 2006 they became Alcatel-Lucent.

This was the first period of growth in a long time for Bell Labs, but it wasn't all good news. By 2008, Alcatel-Lucent announced that it was pulling out of what it called basic science and material research, which included the department that Hendrik once belonged to.

The company was instead going to focus on more marketable areas such as networking software and wireless communications. By July 2008, there were only four on-staff physicists left. The rest had jumped ship for careers in physics departments throughout the US's many distinguished universities. Since then, little has changed.

Nokia acquired Alcatel-Lucent in 2015, and

they've continued operations ever since. Believe it or not, they make a lot more than indestructible phones.

I don't want your takeaway from this to be that economic interests inevitably lead to the death of pure, unfiltered science. There are plenty of teams left in the industry doing amazing research. It's just a lot more hardware-based than it is material science.

If anything, it's remarkable how long Bell Labs managed to stick around as one of the few commercial beacons of science for science's sake.

Wait, did I forget someone earlier?

Bertram Batlogg left Bell Labs for Switzerland long before suspicions are on Hendrik's work became widespread. A quick glance at his publication history tells a fascinating story. You can see his early years toiling away, and then the massive surge due to the superconductor boom of the 80s and 90s. This work would continue to sustain a pretty healthy yearly citation rate, even to this day.

But, towering above all the rest of his accomplishments, is the three-year period where he supervised Jan Hendrik Schön.

Despite the criticisms in the Bell Labs report, he

didn't lose his job in Zurich. He had no titles revoked. He almost certainly would have had a larger research group if things had gone differently, but considering how central he was to the whole ordeal, he did well for himself. If anything, the most visible impact the scandal had on him was that it takes up half his Wikipedia entry.

In the years since the scandal broke, Batlogg has adopted a reasonably humble attitude. I think he's accepted his partial responsibility for the scandal now that some time has passed, and he even became involved in teaching a course on scientific ethics at the Swiss institute for technology.

The month after Hendrik was fired, in the heated aftermath of the scandal, Batlogg is quoted as saying the following:

"When I am a passenger in a car, and the driver drives through a red light, then I am not to blame."

Today, Batlogg would probably disagree with that statement. To borrow from his own analogy:

If you're the passenger advising a reckless and inexperienced driver, it's your job to pull that car to the side of the road.

I really hope that's the closure you wanted, because that's all I have for you.

[RETRACTED]

REFERENCES

Beasley, Malcolm R, Supriyo Datta, Herwig Kogelnik, Herbert Kroemer, and Don Monroe. 2002. REPORT OF THE INVESTIGATION COMMITTEE ON THE POSSIBILITY OF SCIENTIFIC MISCONDUCT IN THE WORK OF HENDRIK SCHÖN AND COAUTHORS. Document. September.

MIT Technology Review. 2002. Innovators Under 35 2002.

Nobel Foundation. 2023. NobelPrize.org.

Reich, Eugene S. 2009. Plastic Fantastic: How the biggest fraud in physics shook the scientific world. Palgrave Macmillan.

Science. 2001. Breakthrough of 2001: Nanoelectronics. December 20.

[RETRACTED]

ABOUT THE AUTHOR

Dylan King is an author and journalist who has contributed to various outlets, including Sports Illustrated, Deadspin, Arrowhead Report, and SB Nation.

He is currently studying Neuroscience and Evolutionary Medicine at Arizona State, where he is a researcher with the Arizona Cancer Evolution Center.

King has two books available from independent publisher Monarch Press:

[RETRACTED]: The Fraud That (Nearly) Won a Nobel Prize, and Section One: When Football Becomes Life or Death, available on Amazon.

[RETRACTED]

www.ingramcontent.com/pod-product-compliance
Lightning Source LLC
Chambersburg PA
CBHW060851220526
45466CB00003B/1333